Glass Fusing for Beginners

Chapter One

Introduction

Glass fusing is a unique and fascinating art form that involves melting pieces of glass together to create beautiful and intricate designs. Whether you are a beginner looking to explore a new hobby or an experienced artist seeking to expand your skill, glass fusing offers endless creative possibilities.

What is Glass Fusing?

Glass fusing is the process of melting two or more pieces of glass together to create a single, solid piece. The glass is heated in a kiln to a temperature that causes it to soften and fuse. Once the glass is fused, it can be shaped, cut, and embellished to create a wide range of designs, from simple bowls and coasters to intricate sculptures and jewelry.

Why learn glass fusing?

There are many reasons why people choose to learn glass fusing. For one, it is a highly creative and rewarding art form that allows you to express yourself in unique and beautiful ways. Glass fusing also offers the opportunity to work with a wide range of colors and textures, making it a versatile medium that can be adapted to suit many different styles and tastes.

In addition to the creative benefits, glass fusing is also a great way to relieve stress and unwind. The process of working with glass can be very meditative and calming, making it an ideal hobby for anyone looking to de-stress and relax.

After this comprehensive introduction to glass fusing for beginners, we're looking to cover the techniques used in going about it, and offer tips for combining glass fusing with other art forms. Whether you're looking to create functional pieces, like plates and bowls, or decorative works of art, This book will provide the knowledge and inspiration you need to get started.

Chapter Two

Materials and Tools

Before you can start creating beautiful fused glass pieces, it's important to have a good understanding of the materials and tools you will need.

Types of Glass Used in Fusing

There are several types of glass used in fusing, including sheet glass, frit, stringers, and rods. Sheet glass is the most commonly used type of glass and comes in a wide range of colors and textures. Frit is crushed glass that comes in a variety of sizes, from fine powder to coarse granules. Stringers are thin, straight pieces of glass that can be used to create fine lines and details, while rods are thicker pieces of glass that can be used for sculpting and shaping.

Basic Tools and Equipment

To get started with glass fusing, you will need a few basic tools and equipment.

These include a glass cutter, running pliers, breaking pliers, a grinder, and a kiln. You will also need safety equipment, such as safety glasses and heat-resistant gloves, to protect yourself during the fusing process.

Safety Precautions

Working with glass can be dangerous if proper safety precautions are not taken. It's important to always wear safety glasses when cutting or grinding glass and to use heat-resistant gloves when handling hot glass. You should also be sure to properly ventilate your workspace and keep flammable materials away from your kiln.

By taking the necessary safety precautions and using the right tools and equipment, you can create a beautiful fused glass piece with confidence and peace of mind.

Chapter Three

Basic Techniques of Glass Fusing

Cutting and Shaping Glass

The first step in glass fusing is to cut and shape your glass pieces. This can be done using a glass cutter and running pliers, which allow you to break the glass along the score line. You can also use a grinder to shape your glass pieces and smooth any rough edges.

Creating a Basic Fused Glass Piece

Once you have cut and shaped your glass piece, you can start creating a basic fused glass piece. This can be done by layering your glass pieces on top of each other and then placing them in the kiln to fuse. This process typically takes several hours, as the glass needs to heat up slowly and cool down gradually to prevent cracking.

Layering Glass for More Complex Designs

Once you have mastered the basic fused glass techniques, you can start experimenting with layering glass for more complex designs. This can be done by adding different colors and textures of glass to your piece, as well as incorporating other materials, such as metals or ceramics.

By layering your glass in creative and unique ways, you can create stunning works of art that are truly one-of-a-kind.

Chapter Four

Advanced Techniques and Design

Slumping and Draping

Slumping and draping are advanced techniques of glass fusing that involve shaping the glass into a specific form.

Slumping involves heating the glass in a mold until it softens and conforms to the shape of the mold. Draplng is similar but involves heating the glass until it slumps over a mold, creating a curved or rounded shape.

Designing Your Fused Glass Pieces

Design is an important aspect of glass fusing, as it allows you to create truly unique pieces that reflect your style. You can experiment with different colors, textures, and shapes of glass to create interesting and dynamic designs. You can also incorporate other materials, such as metals or ceramics, to add visual interest to your pieces.

Decorating Your Fused Glass Pieces

Once you've created your fusing glass piece, you can decorate it in a variety of ways. This can be done using techniques such as sandblasting, engraving, or painting. You can also add decorative elements such as mosaics, dichroic glass, or glass beads to your piece to create a truly stunning work of art.

By combining advanced techniques and decoration, you can create fused glass pieces that are truly unique and showcase your skills as a glass artist.

Chapter Five: Maintaining and Caring for Your Fused Glass Pieces

Handling and Displaying Your Fused Glass Pieces

One of the most important aspects of maintaining your fused glass pieces is proper handling and display. You should always handle your fused glass pieces with care, using soft, lint-free cloths or glows to prevent smudges or scratches. When displaying your pieces, you should avoid direct sunlight and excessive heat or moisture, which can damage the glass over time.

Cleaning Your Fused Glass Places

Regular cleaning is also important for maintaining the beauty of your fused glass pieces. You should clean your pieces using a soft, lint-free cloth and a gentle, non-abrasive cleaner. Avoid harsh chemicals or abrasive materials that can stretch or damage the glass.

Repairing and Restoring Your Focus Glass Pieces

Despite your best efforts, your fused glass pieces may occasionally need repair or restoration. This can be done using techniques such as coldworking, which involves grinding or polishing the glass to remove scratches or clips, or using a UV-curing adhesive to reattach broken pieces.

By taking care of your fused glass pieces and addressing any repairs or restorations as needed, you can ensure that your pieces continue to be enjoyed and appreciated for many years to come.

Chapter Six

Resources for Glass-Fusing Enthusiasts

Books and Magazines

There are many excellent books and magazines available on the topic of glass fusing. These resources can provide inspiration, techniques, and tips for creating beautiful fused glass pieces. Some recommended titles include **"Introduction to Glass Fusing"** by **Petra Kaiser**, **"Glass Fusing Basics"** by **Lynn Haunstein**, and **"Glass Art Magazine"**.

Website and Online Communities

The internet is a great resource for glass fusing enthusiasts. There are many websites and online communities dedicated to the art of glass fusing, where you can connect with other artists, find inspiration, and get tips and advice. Some popular websites and communities include **Glass Campus**, **Glass Art Society**, and **Glass Artisans**.

Local Classes and Workshops

Taking classes and workshops is an excellent way to improve your skills and connect with other glass-fusing enthusiasts in your area. Many community centers, art studios, and glass supply stores offer classes and workshops on glass fusing, where you can learn new techniques and get hands-on experience. You can also connect with local glass fusing groups or clubs to meet other enthusiasts and participate in group activities.

Conclusion

Glass fusing is a rewarding and enjoyable hobby that allows you to create beautiful works of art using glass. By mastering basic techniques and experimenting with more advanced techniques and designs, you can create unique and stunning pieces that reflect your style and creativity. With proper care and maintenance, your fused glass pieces can continue to be enjoyed for many years to come.

And for beginners, utilizing resources available to you, like books, websites, online communities, and local classes, you can continue to develop your skills and connect with other glass fusing enthusiasts. With dedication and practice, you can become a skilled glass-fusing artist and create beautiful works of art that are sure to impress.